PLATE TECTONICS

EILEEN GREER

PowerKiDS press™

NEW YORK

Published in 2017 by The Rosen Publishing Group, Inc.
29 East 21st Street, New York, NY 10010

Editor: Melissa Raé Shofner
Book Design: Michael Flynn
Interior Layout: Mickey Harmon

Cataloging-in-Publication Data
Names: Greer, Eileen.
Title: Plate tectonics / Eileen Greer.
Description: New York : PowerKids Press, 2017. | Series: Spotlight on earth science| Includes index.
Identifiers: ISBN 9781499425321 (pbk.) | ISBN 9781499425352 (library bound) | ISBN 9781499425338 (6 pack)
Subjects: LCSH: Plate tectonics--Juvenile literature.
Classification: LCC QE511.4 G74 2017 | DDC 551.1'36--dc23

Manufactured in the United States of America

CPSIA Compliance Information: Batch #BW17PK For further information contact Rosen Publishing, New York, New York at 1-800-237-9932.

CONTENTS

CHAPTER ONE

WHAT IS PLATE TECTONICS?

You might think that Earth stays pretty much the same over time. However, the surface of Earth is constantly in motion. Over millions of years, mountains and volcanoes rise higher and **trenches** sink deeper. These things happen because of plate tectonics, which is the idea that Earth's top layer is broken into pieces called plates. The layer below the plates is very different. It allows the plates to drift away from each other and bump into each other.

The first person to suggest that Earth's top layer could move was Alfred Wegener. He noticed that the continents seemed to fit together like puzzle pieces. He proposed that Earth once had one large continent and that pieces of it broke away over a long period of time. Many people were against Wegener's idea of continental drift. It wasn't until the 1960s that people started to understand plate tectonics.

ALFRED WEGENER

You can stand on the boundary between the North American plate and Eurasian plate at Thingvellir National Park in Iceland.

CHAPTER TWO

EARTH'S LAYERS

Earth isn't one solid ball floating in space. The planet is actually made up of several different layers. The deeper the layer, the greater the heat and pressure become. The innermost layer is the solid inner **core**. This area has the most pressure and heat of all the layers. The next layer is the outer core, which is liquid.

The following layer is the mantle. It's estimated to be around 1,800 miles (2,900 km) thick. It's made up of many elements, including iron, calcium, and magnesium. The mantle has a few parts. Earth's top layer is the crust. The crust is the thinnest layer. You're probably most familiar with Earth's crust because every landform you've ever seen is a part of it. The top part of the mantle and the crust together are called the lithosphere. This is the layer made up of tectonic plates.

The asthenosphere is made of hot, soft rock. Tectonic plates "float" on the asthenosphere.

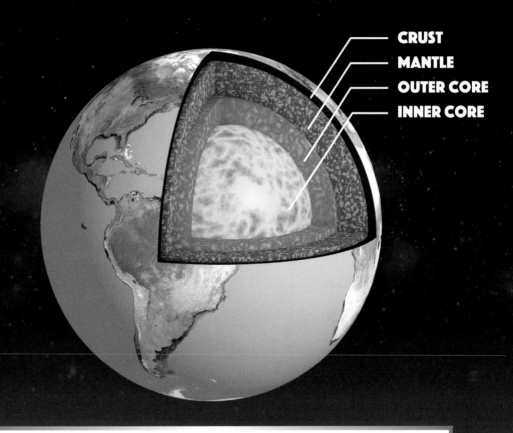

CRUST
MANTLE
OUTER CORE
INNER CORE

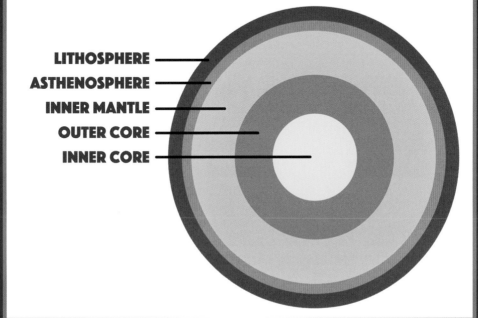

LITHOSPHERE
ASTHENOSPHERE
INNER MANTLE
OUTER CORE
INNER CORE

CHAPTER THREE

GO WITH THE FLOW

How do the plates move? They go with the flow! This flow starts deep inside Earth where it's very hot. Magma from the core rises through the mantle toward Earth's surface, and heat rises with it. The heat causes the rock in the asthenosphere to act somewhat like melted plastic. Heat forces the rock up to the lithosphere, where much of the heat is released through the crust. The rock in the asthenosphere then cools and flows back toward Earth's center, where it will eventually heat up again. This flow of heat inside the asthenosphere is called a **convection** current.

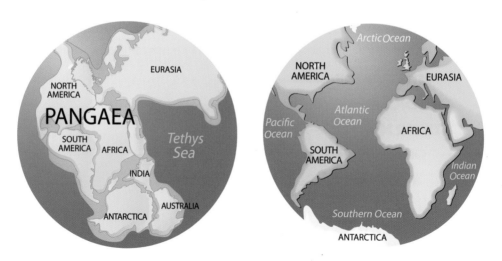

Alfred Wegener first suggested the idea of Pangaea in the early 1900s. Pangaea was a giant continent that existed millions of years ago. He believed that plate tectonics had caused Pangaea to break up and drift apart into seven separate continents.

Convection currents cause tectonic plates to drift. Some come together, while others pull apart. Still others grind past each other. All this movement has a great effect on the shape of Earth's land.

CONVECTION CURRENTS

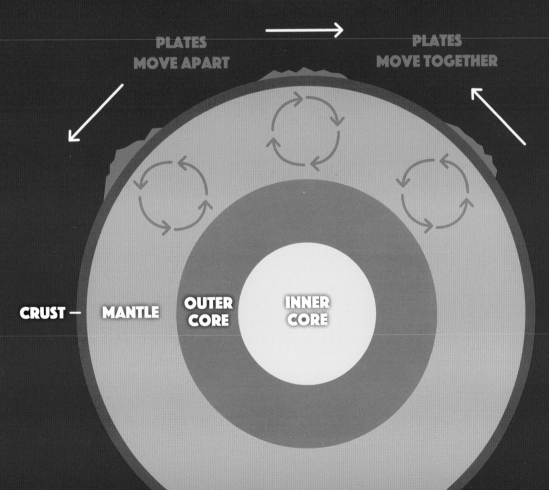

PLATES MOVE APART

PLATES MOVE TOGETHER

CRUST — MANTLE OUTER CORE INNER CORE

CHAPTER FOUR

EARTH'S MANY PLATES

There are two kinds of crust on Earth. Continental crust is found under the continents. Oceanic crust is found under the oceans. Oceanic crust is often younger and thinner than continental crust.

Plate movement sometimes causes volcanoes to erupt. **Magma** under pressure is released and explodes onto Earth's surface as lava.

PACIFIC PLATE

The Pacific plate is the largest tectonic plate on Earth.

The plates that make up Earth's crust have names, which often indicate their continent, ocean, or region. Tectonic plates are made of both oceanic crust and continental crust. The largest plates are the North American plate, South American plate, Eurasian plate, African plate, Pacific plate, Australian plate, and Antarctic plate. Smaller plates include the Nazca plate, Indian plate, Arabian plate, Philippine plate, and Caribbean plate.

There are many mountains, volcanoes, and trenches along the boundaries between these plates. The Pacific Ring of Fire is an area with hundreds of volcanoes and frequent earthquakes. It's located where the Pacific plate meets other plates such as the North American plate and Eurasian plate.

CHAPTER FIVE

DIVERGENT BOUNDARIES

When two plates pull apart from each other, it's called a divergent boundary. If plates pull apart on land, it causes a trough. Troughs are long, narrow **depressions** in the landscape. One example of a trough created by a divergent boundary is the Great Rift Valley in Africa and Asia. The rift is like a fracture, or break, in Earth's surface. It gets wider over time. Scientists believe that one day, east Africa may split apart from the rest of the continent because of this divergent boundary.

GREAT RIFT VALLEY

SEAFLOOR SPREADING

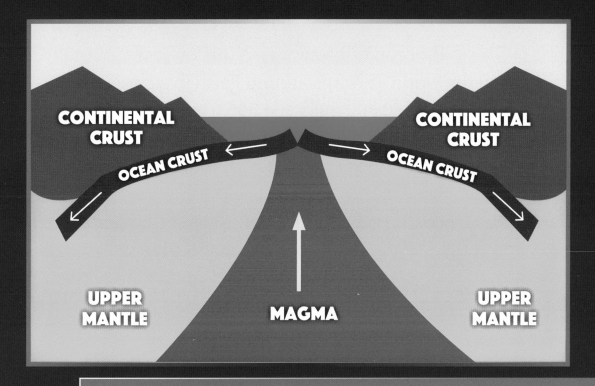

CONTINENTAL CRUST

OCEAN CRUST

CONTINENTAL CRUST

OCEAN CRUST

UPPER MANTLE

MAGMA

UPPER MANTLE

"Seafloor spreading" is a term used to describe what happens at divergent boundaries in the ocean. New oceanic crust is formed when magma rises to the seafloor and continents are pushed farther away from one another.

Divergent boundaries also happen in the ocean. When the plates pull apart, magma rises from the mantle to the ocean floor. The magma hardens and cools. Over time, it may build up into mountains or volcanoes. That means that the ocean floor is always being recreated. The moving plates also widen the ocean's **basins**.

CHAPTER SIX

MID-OCEAN RIDGE

The mid-ocean ridge system is the longest mountain range in the world. It wraps around Earth in a curve that's more than 40,300 miles (64,856.6 km) long. Most of it is underwater, and it was formed by diverging tectonic plates. It's made up of ocean ridges such as the Mid-Atlantic Ridge and the East Pacific Rise. The Mid-Atlantic Ridge stretches from northern polar waters to southern polar waters.

The tectonic plates that are responsible for the mid-ocean ridge system are still spreading apart. They can move less than an inch (2.5 cm) or up to 8 inches (20.3 cm) each year. This adds to the long line of mountains and volcanoes in the ridge. These volcanoes erupt as magma rises to the surface. Most of the time, we don't notice when an underwater eruption occurs because it's so far under the ocean's surface.

Islands such as Iceland and Bouvet Island, Norway, are parts of the Mid-Atlantic Ridge that rise above the water's surface.

BOUVET ISLAND, NORWAY

ICELAND

CHAPTER SEVEN

CONVERGENT BOUNDARIES

Convergent boundaries are places where two tectonic plates collide, or run into each other. At the collision point, the crust moves upward. This forms mountain ranges. The mountain ranges grow as time goes on. Mount Everest is the tallest mountain above **sea level** in the world. It's part of the Himalayas, a mountain range that formed when two plates collided more than 55 million years ago.

Some convergent boundaries form trenches. Trenches are created when an ocean plate meets a landmass. The ocean plate dives underneath the landmass, making a deep gap. The deepest trench in the world is the Mariana Trench. Its deepest point, the Challenger Deep, is nearly 7 miles (11.3 km) beneath the ocean's surface. As the oceanic plate dives, the other plate's landmass rises. Some of the Andes mountain range in South America formed where the Nazca plate dove under the South American plate.

MOUNT FUJI

JAPAN

When two oceanic plates collide, one often dives under the other. This can cause volcanoes to form as parts of the lower plate melt and the magma rises. The islands of Japan were formed this way.

CHAPTER EIGHT

TRANSFORM BOUNDARIES

Transform boundaries are places where two plates slip past one another. Plates aren't smooth, and they don't slip gracefully. Their jagged edges grind against one another. This motion doesn't form mountains, trenches, or volcanoes. However, the energy released at these points causes seismic activity, or earthquakes.

The Pacific plate and North American plate form a transform boundary at the San Andreas Fault zone in California. Faults are cracks in Earth's crust. They're found along plate boundaries, but they can form elsewhere, too. There are dozens of faults across California, but the San Andreas Fault is the most well known. In 1906, an earthquake **originating** from the San Andreas Fault destroyed most of San Francisco. The earthquake caused a massive fire and killed as many as 3,000 people. Scientists warn that it is only a matter of time before the San Andreas Fault sends another huge earthquake ripping through the earth.

The Alpine Fault in New Zealand is another example of a transform boundary.

CHAPTER NINE

EARTHQUAKES

The three main kinds of faults are normal, **reverse**, and strike-slip. A normal fault occurs when a piece of crust moves downward and away from another piece. When a piece of crust moves upward and toward another piece, that causes a reverse fault. A strike-slip fault happens when two pieces of crust move side by side in opposite directions. Some faults are more than 1,000 miles (1,609 km) long, while others are less than an inch (2.5 cm) long.

The **friction** at a fault can build up for hundreds of years. When the plates finally slip and release the energy, it may cause a big earthquake. The fault sends energy through the earth in waves. Primary waves, or P waves, push and pull their way through the earth. Secondary waves, or S waves, move up and down or from side to side. After the **initial** earthquake, other **tremors** may be felt for days or even weeks afterwards. They're called aftershocks.

Scientists who study earthquakes are called seismologists. They measure the strength of earthquakes and try to predict when one might occur. Unfortunately, it's nearly impossible to predict an earthquake.

CHAPTER TEN

EVER-CHANGING EARTH

Scientists believe Earth is around 4.5 billion years old. Earth has changed a lot over that time. Landmasses have formed, continents have moved, and the ocean floor has been created and recreated. This is all thanks to plate tectonics.

Where can you go to observe the results of plate tectonics? You can visit Thingvellir National Park in Iceland to walk between the North American and Eurasian plates. You can also visit mountain chains such as the Rocky Mountains in the western United States or faults such as the San Andreas Fault in California. These landforms show us how shifting plates have changed Earth and how it might change more in the future.

Inch by inch, plates move together, apart, or along one another. Their movement shapes our world. Earth is truly an ever-changing planet.

THE ROCKY MOUNTAINS

GLOSSARY

basin (BAY-suhn) A dip in the surface of Earth occupied by an ocean or another body of water.

convection (kuhn-VEK-shun) A motion in a fluid or gas in which the hot parts rise and the cold parts sink.

core (KOR) The center of something.

depression (dih-PREH-shun) An area of Earth's surface lower than the area around it.

friction (FRIK-shun) The force that resists motion between bodies in contact.

initial (ih-NIH-shul) Relating to the beginning.

magma (MAG-muh) Hot, liquid rock inside Earth.

originate (uh-RIH-juh-nayt) To start from or cause the beginning of.

reverse (rih-VERS) Opposite to what is usual or normal.

sea level (SEE LEH-vuhl) The average height of the sea's surface.

tremor (TREH-muhr) A shaking movement of the ground caused by an earthquake.

trench (TRENCH) A long, narrow cut in the ground.

INDEX

PRIMARY SOURCE LIST

Page 4
Alfred Wegener. Photograph. ca. 1924–1930. From Bildarchiv Foto Marburg Aufnahme-Nr. 426.294.

Page 15
Bouvet Island, southeast side. Black-and-white photograph. By Carl Chun. Colored by F. Winter. Taken on German Valdivia expedition. November 26, 1898. From NOAA Photo Library.

Page 17
Satellite image of Japan. May 1, 2003. By Jeff Schmaltz, MODIS Rapid Response Team, NASA/GSFC. From NASA.

WEBSITES

Due to the changing nature of Internet links, PowerKids Press has developed an online list of websites related to the subject of this book. This site is updated regularly. Please use this link to access the list: www.powerkidslinks.com/soes/plate